ARMY RANGERS

JIM WHITING

CREATIVE EDUCATION • CREATIVE PAPERBACKS

PUBLISHED BY Creative Education and Creative Paperbacks
P.O. Box 227, Mankato, Minnesota 56002
Creative Education and Creative Paperbacks are imprints of
The Creative Company
www.thecreativecompany.us

DESIGN BY Christine Vanderbeek; **PRODUCTION BY** Liddy Walseth
ART DIRECTION BY Rita Marshall
PRINTED IN CHINA

PHOTOGRAPHS BY
Alamy (501 collection, AF archive, Chronicle, DOD Photo, Entertainment Pictures, Moviestore collection Ltd, National Geographic Creative, North Wind Picture Archives, PJF Military Collection, RGR Collection, Science History Images, Stocktrek Images, Inc., Sueddeutsche Zeitung Photo, United Archives GmbH, Z2A1), DVIDS (SGT Coty Kuhn, SPC David Shefchuk), iStockphoto (spxChrome), Shutterstock (ALMAGAMI, Getmilitaryphotos, gst)

COPYRIGHT © 2019 CREATIVE EDUCATION,
CREATIVE PAPERBACKS
International copyright reserved in all countries. No part of this book may be reproduced in any form without written permission from the publisher.

LIBRARY OF CONGRESS CATALOGING-IN-PUBLICATION DATA
Names: Whiting, Jim, author.
Title: Army Rangers / Jim Whiting.
Series: U.S. Special Forces.
Includes bibliographical references and index.
Summary: A chronological account of the American military special forces unit known as the Army Rangers, including key details about important figures, landmark missions, and controversies.
Identifiers: LCCN 2017028017 / ISBN 978-1-60818-983-0 (hardcover) / ISBN 978-1-62832-610-9 (pbk) / ISBN 978-1-64000-084-1 (eBook)
Subjects: LCSH: 1. United States. Army. Ranger Regiment, 75th—Juvenile literature. 2. United States. Army—Commando troops—Juvenile literature.
Classification: LCC UA34.R36 W55 2017 / DDC 356/.1670973—dc23

CCSS: RI.5.1, 2, 3, 8; RH.6-8.4, 5, 6, 8

FIRST EDITION HC 9 8 7 6 5 4 3 2 1
FIRST EDITION PBK 9 8 7 6 5 4 3 2 1

TABLE OF CONTENTS

INTRODUCTION **9**
FROM COLONIAL AMERICA TO AFGHANISTAN **11**
SOME OF THE WORLD'S TOUGHEST TRAINING **19**
FIREPOWER AND FILMS **27**
RANGERS IN ACTION **35**

★ ★ ★

GLOSSARY **45**
SELECTED BIBLIOGRAPHY **47**
WEBSITES **47**
READ MORE **47**
INDEX **48**

Army Rangers are known for their ability to move without being seen by their enemies.

FORCE FACTS In their ongoing training, Rangers focus on the Big Five: small-unit tactics, mobility, marksmanship, physical training, and medical training.

FORCE FACTS At Pointe du Hoc, one Ranger climbed to the top of an extension ladder mounted on a landing craft. As it swayed, he fired his weapon at the top of the arc and then held on.

INTRODUCTION

American, British, and Canadian forces planned to invade the beaches of Normandy, France, on June 6, 1944, in an effort to end World War II in Europe. But surveillance showed six massive German guns perched atop a cliff at Pointe du Hoc. From that vantage point, they could rain down heavy shells on the troops coming ashore and warships lying offshore. A group of 225 United States Army Rangers was assigned the task of destroying the guns. To do so, they would have to scale a cliff more than 100 feet (30.5 m) high while taking fire from above. One *intelligence* officer said, "It can't be done. Three old women with brooms could keep the Rangers from climbing that cliff."

The Rangers arrived at the foot of the cliff later than planned. German defenders had time to recover from the intense bombardment delivered by Allied aircraft and warships. Some Rangers had to wade ashore in frigid, shoulder-deep water. The ropes attached to their *grappling hooks* were waterlogged and heavy. That made them harder to throw.

Still, the men scrambled to the top of the cliff, where they made a startling discovery. The "guns" were actually heavy logs that had fooled air *reconnaissance*. The troops located and disabled the real weapons. The Germans rushed mobile artillery to the position. Outnumbered and outgunned, the Rangers held on for 48 hours amidst fierce counterattacks until reinforcements finally arrived. "The Rangers at Pointe du Hoc were the first American forces on D-Day to accomplish their mission, and we are proud of that," said one of the survivors.

Before Pointe du Hoc, Rangers practiced climbing cliffs along the coast of England.

FROM COLONIAL AMERICA TO AFGHANISTAN

THE RANGERS' HISTORY IS ROOTED IN AMERICA'S. Early European settlers quickly discovered that American Indians resisted intrusions on their land. They didn't "fight fair" by European standards of warfare. "They never fight in open fields but always among reeds or trees, taking their opportunity to shoot at their enemies till they can knock [notch] another arrow they make the trees the defense," said Captain John Smith soon after helping to establish the colony of Jamestown, Virginia, in 1607. Colonists began forming militia units to defend against attacks. Some became especially skilled in "ranging" through the wilderness from one settlement to another, looking for signs of impending danger.

In 1675, Colonel Benjamin Church of Massachusetts took "ranging" a step farther. He organized the Plymouth Colony Militia. It was a mixture of friendly American Indians and colonists with woodland skills. The force launched attacks during conflicts such as King Philip's War (1675–76) and King William's War (1689–97). In 1716, Church's *Entertaining Passages Relating to Philip's War* was published. Historians regard it as the first American military manual.

The first significant unit to call themselves "rangers" was Rogers' Rangers. They were organized and led by Major Robert Rogers during the *French and Indian War* (1754–63). Many authorities credit Rogers with creating the techniques of Long

Early Rangers wore green and other drab-colored uniforms to conceal themselves in the woods.

> **FORCE FACTS** Robert Rogers often paid his men out of his own pocket. Neither the British nor colonial governments ever repaid him.

Range Reconnaissance Patrol (LRRP). His 28 "Rules of Ranging" combined his own combat techniques with those he picked up from both his American Indian allies and his enemies. He was especially adept at operating during the wintertime, when traditional military units were unable to function. At war's end, Rogers' Rangers disbanded.

Rogers offered his services to George Washington at the outset of the Revolutionary War. When Washington turned him down, Rogers fought for the British. Several Revolutionary leaders established units that performed many of the Rangers' functions. One of the best known was Francis Marion. The British called him the "Swamp Fox." His men launched hit-and-run attacks from the swamps and dense forests of South Carolina. Then they returned to their wildland sanctuary, where the British were unable to follow.

As the new country expanded its territory westward during the 1800s, unofficial mounted militia units performed Ranger-like scouting missions. The Civil War marked the return of formal ranger military units. Probably the best known was Mosby's Rangers. The group was organized and led by the "Gray Ghost," Confederate Colonel John S. Mosby. His men were noted for mounting quick attacks against Union forces before melting back into the countryside.

The Rangers were revived in the early days of World War II to fill a need for fast-striking, mobile units. The new group's commander was Lieutenant Colonel William O. Darby. One of Darby's first acts was reading Rogers' "Rules of Ranging" to his men. The Rangers acquired their enduring motto during the Normandy assault. With hundreds of troops killed by German gunfire and thousands more pinned down, Brigadier General Norman Cota reportedly shouted, "Rangers lead the way!" Individual acts of bravery by Rangers and other soldiers soon got the men off the beach, and the invasion became successful.

Before that, however, the Rangers had suffered a significant

ABOVE: During the Civil War, Union officers repeatedly tried and failed to capture Mosby.

> **FORCE FACTS** Merrill's Marauders went into action in February 1944. The unit was disbanded six months later. At that point, only 130 of the original 2,997 men remained combat-ready.

setback at Italy's Battle of Cisterna in January 1944. Given faulty intelligence, lightly armed Rangers attacked the village of Cisterna. They expected to encounter little opposition. Unfortunately, the Germans had tanks and large numbers of troops waiting. Of the 767 Rangers who began the attack, only 6 returned safely. The others were killed or captured.

Rangers also served in the Pacific. The Sixth Ranger Battalion rescued more than 500 prisoners in a daring raid in the Philippines. The 5307th Composite Unit (Provisional), better known as Merrill's Marauders, was led by General Frank Merrill. This group conducted a campaign in the jungles of Burma (now Myanmar). The unit marched more than 1,000 miles (1,609 km) without adequate food, in stifling temperatures and humidity, and contending with clouds of mosquitoes and serious tropical diseases. Every man received the Bronze Star after the unit's final victory at the Battle of Myitkyina in August 1944.

The Rangers were disbanded after the war, only to be called back into action at the outbreak of the Korean War in 1950. Seventeen Ranger companies were formed. Each was attached to a larger army unit to provide scouting and *small-unit tactics*. The Army also established the Ranger School, a two-month period of intensive training. When that war ended in 1953, the Rangers were once again disbanded. The school remained in use because the army felt its graduates would upgrade the performance of the units they were assigned to.

In the early 1960s, the army formed LRRP companies in Europe to provide early warnings of Soviet attacks during the Cold War. During the Vietnam War, LRRP companies were attached to every American brigade

Merrill's Marauders trudged hundreds of miles through Burma during World War II.

Marching in straight rows, Rangers keep enough space between them to prevent one bullet from killing two men.

and division. As conflict in Vietnam wound down, the Rangers weren't disbanded. Instead, the army chief of staff, Creighton Abrams, ordered the formation of First Ranger Battalion in January 1974. Six months later, Second Battalion was formed. In 1984, Third Battalion and a headquarters component rounded out what soon became known as the Seventy-Fifth Ranger Regiment, the force's official name.

By then, the Rangers had become well-established as an elite Special Forces branch. Elements of First Battalion participated in Operation Eagle Claw, the attempt to rescue American hostages in Iran in 1980. Lack of communication among participating units, coupled with crucial equipment failures, doomed the mission. Three years later, both First and Second Battalions played key roles in Operation Urgent Fury. A communist government had taken over the Caribbean island of Grenada. It began building an airstrip. U.S. president Ronald Reagan feared that the airstrip could be used for military purposes. He was also concerned about the safety of hundreds of American medical students on the island and the possibility of another hostage situation.

The Rangers conducted an airborne assault that helped take down the communist government. In December 1989, all three battalions took part in Operation Just Cause, which overthrew Panamanian dictator Manuel Noriega. Noriega was brought to the U.S. to stand trial on drug trafficking charges.

The Rangers' next major operation took place in Somalia in 1993. The Rangers provided security coverage for several **Delta Force** operations that sought to apprehend warlords who were interfering with humanitarian relief efforts. One operation, now known as the Battle of Mogadishu, turned disastrous when it fell behind schedule. Thousands of armed Somali fighters swarmed the area and laid down heavy fire. This battle was later called the most intense ground combat since the Vietnam era. Several Rangers were killed and many were wounded. The resulting chaos led to American withdrawal shortly afterward.

After the September 11, 2001, terrorist attacks in the U.S., the Rangers went to work again. They played a key role in overthrowing the **Taliban** in Afghanistan. The Taliban had provided training sites for the 9/11 terror attackers. In March 2003, Rangers conducted the first airborne assault in Iraq as the War on Terror expanded there.

In 2007, the Seventy-Fifth Regimental Special Troops Battalion (RSTB) was added to the Regiment to provide support in several key areas. One component is the Regimental Reconnaissance Company, or RRC. It carries out special recon behind enemy lines in preparation for action by a larger Ranger force. RRC may also direct air strikes, secure drop zones, and install equipment such as navigation beacons. The company consists of an unknown number of six-man teams. Other RSTB companies deal with communications, coordinating intelligence gathering, and the selection and training of future Rangers.

The Rangers have become one of the few American combat units continuously in action during the War on Terror. The force conducts a variety of operations in Afghanistan and Iraq and is ready to go anywhere it is needed.

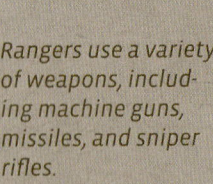

Rangers use a variety of weapons, including machine guns, missiles, and sniper rifles.

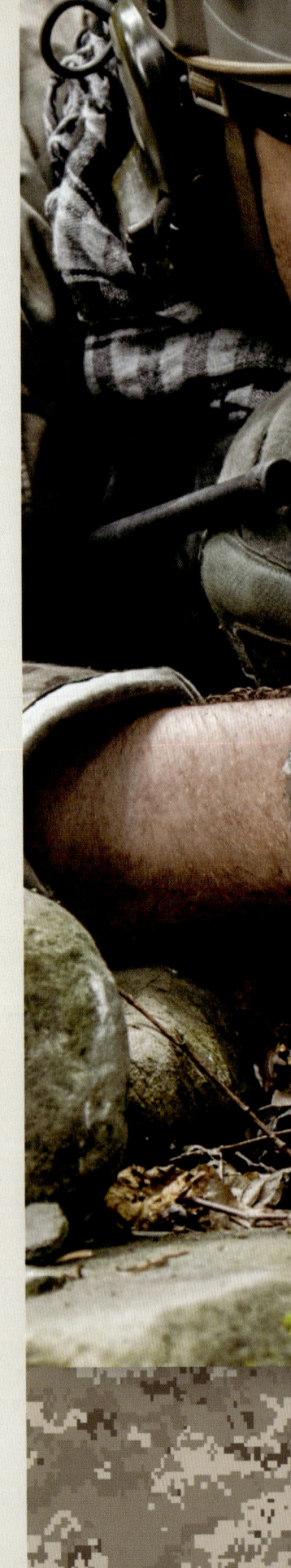

FORCE FACTS Originally, Ranger battalions had three rifle companies. The fourth battalion, RSTB, was added in 2007 because of the increasing number of deployments at that time.

SOME OF THE WORLD'S TOUGHEST TRAINING

Today, Rangers are best known for, and most used in, direct-action missions. According to the U.S. Department of Defense, direct action consists of "short-duration strikes and other small-scale offensive actions conducted as a special operation in hostile, denied, or politically sensitive environments and which employ specialized military capabilities to seize, destroy, capture, exploit, recover, or damage designated targets. Direct action differs from conventional offensive actions in the level of physical and political risk, operational techniques, and the degree of discriminate and precise use of force to achieve specific objectives."

One example of Rangers' direct action is conducting airborne raids to seize and hold airfields in preparation for the arrival of conventional forces. Other examples include small-unit attacks to capture high-value targets (HVTs), demolitions, hostage rescues, and equipment recovery. Rangers usually *deploy* in platoon (about 40 men) or company (about 150 men) strength. When high-profile situations demand it, though, one of the battalions or even the entire regiment will go into action as a unit.

To carry out such dangerous missions, Rangers need to be highly trained and in outstanding physical condition. Not surprisingly, therefore, becoming a Ranger is a difficult process. To be considered, candidates must volunteer and be U.S. citizens of good character without physical limitations. They must qualify for Airborne training, pass a psychological screening, be eligible

After numerous practice jumps, Rangers parachute into hostile territory over deserts, jungles, and bodies of water.

> **FORCE FACTS** According to reliable estimates, 75 percent of the operatives in the super-secret Delta Force are recruited from the Seventy-Fifth Ranger Regiment.

for security clearance, and enlist in or already have one Ranger-related Military Occupational Specialty (MOS). The most common MOS for Rangers is infantryman. Others range from maintenance specialties and analysts to administrative assistants and food service operations. There are even positions for chaplain's assistants.

Once accepted, candidates must complete Phase 1 of the Ranger Assessment and Selection Program (RASP 1). Experienced soldiers ranked private through sergeant can begin RASP 1 right away. New recruits must first complete Basic Combat Training and Advanced Individual Training (BCI/AIT). Nearly all candidates complete the three-week Airborne School either before or immediately after RASP 1.

RASP 1 is divided into two four-week phases. Phase 1 involves assessing the potential Ranger's physical and psychological ability to undergo rigorous training. Training includes proficiency in standard Ranger personal weapons. Troops also learn how to use standard enemy weapons, such as the AK-47, in case they recover one on the battlefield. They learn Ranger history and traditions in addition to first aid. Phase 2 is preparation for combat. It includes advanced marksmanship and close-quarters combat (CQC) and close-quarters battle (CQB) training, as well as mobility training using Ranger assault vehicles. Trainees also learn how to *breach* using both handheld and explosive devices.

RASP 2 is a 21-day course at Fort Benning, Georgia. This training is for enlisted ranks of staff sergeant and above who want to become part of the Seventy-Fifth Ranger Regiment. It assesses the suitability of these troops—who already have significant experience as soldiers—to become Rangers and schools them in how Rangers conduct their operations.

Upon completion of the RASP phases, new Rangers receive tan berets. These caps mark them as members of one of America's most elite units. The tan color is in honor of the leather headgear worn by the earliest Rangers.

The Rules of Ranging *advises soldiers to be awake and aware of their surroundings before dawn.*

> **FORCE FACTS** In August 2015, two women became the first females to complete Ranger School. "These two soldiers have absolutely earned the respect of every Ranger instructor," said one of the school's leaders.

Because of the high demand for Ranger operations, new Rangers are almost immediately assigned to one of the three battalions. The battalions consist of a smaller headquarters company and 4 rifle companies of about 150 soldiers each. At the heart of every rifle company are three rifle platoons. These platoons consist of three nine-man squads, each headed by a squad leader. The other eight men are divided into two four-man fire teams. Each platoon also has a seven-man machine gun squad. The company has a weapons platoon, consisting of mortar and anti-tank sections. Through the 1990s, the weapons platoon also included a sniper section. Since then, snipers have been reorganized into a platoon at the battalion level. Depending on where they are deployed, Ranger companies and platoons may have additional personnel. For example, on missions in the Middle East, Rangers are likely to work with a canine team, interpreters, and forward observers to direct heavy weapons fire. They might also employ a Cultural Support Team that knows local customs to ease communication and interactions with local people.

RASP is hardly the end of Ranger training. The troops constantly hone their skills. They often embark on three- or four-day extended simulated training missions, have a couple of days off, and then resume training.

To become Rangers, soldiers must complete several phases of strenuous physical training.

Another option for further training is Ranger School. It has been described by GoArmy.com as "the most physically and mentally demanding leadership school the Army has to offer," and a few members of other services, such as Navy SEALs, may participate. Its primary objective is to develop advanced leadership skills under conditions that stress candidates to the limits of their physical and mental abilities. They undergo more than a

thousand hours of training in 61 days. Many of these hours are at night. The failure rate is high—often more than half.

Ranger School has three phases. First is the Benning (formerly known as Crawl) Phase, which takes place at Fort Benning. The 20 days are divided into 2 parts. First is the Ranger Assessment Phase, also known as RAP week. It opens with the Ranger Physical Assessment. Fitness tests include doing 58 pushups and 69 sit-ups—each in 2 minutes, plus 6 pull-ups, and running 5 miles (8 km) in 40 minutes or less. Then, during the water survival assessment, trainees must cross a log 35 feet (10.7 m) above a pond, crawl along a rope, and then drop into the water. There are also daytime and nighttime navigation tests. RAP week concludes with a 12-mile (19.3 km) march. Trainees are not allowed to drink water and must complete the march in under three hours while carrying a weapon and wearing a 35-pound (16 kg) *rucksack*. The second part of the Benning Phase is called the patrolling phase. Students learn all aspects of carrying out a successful mission. It includes the Darby Queen, a 20-obstacle course

Captain Kristen Griest and First Lieutenant Shaye Haver were the first women to earn the Ranger tab.

set over a mile (1.6 km) of hilly terrain, and it concludes with peer evaluation. This ensures that those who will move on to the next phase have the full confidence of their fellow soldiers.

That next phase, the Mountain (previously Walk) Phase, takes place at Camp Frank D. Merrill in the mountains of northern Georgia. In 21 days, students learn all aspects of mountaineering techniques and patrolling. They put this newfound knowledge to good use in multi-day simulated patrols and attacks on "enemy" forces under conditions of severe weather, hunger, sleep deprivation, and emotional and physical stress. Part of the mental stress comes from a profound sense of isolation. The camp is miles from civilization. Some of the physical stress is caused by low nighttime temperatures, especially during winter.

The final segment—the Florida (previously Run) Phase—takes place at Camp James E. Rudder at Elgin Air Force Base in Florida. This phase trains soldiers to function effectively in swamp environments. It also teaches methods of leading and coordinating airborne, small-boat, and ship-to-shore assaults. At the completion of this phase, soldiers receive the Ranger tab, which becomes part of their uniform for the rest of their military careers.

Other specialized training programs include: the Ranger radio telephone operator (RTO) and Pre-Special Operations Combat Medic courses (before the Special Operations Combat Medic course); Ranger Language Program; Military Free-Fall (parachute) School; Scuba School; Survival, Evasion, Resistance, and Escape (SERE) training; and Military Operations on Urban Terrain (MOUT) training. Such grueling training regimens produce one of the world's finest light infantry forces. The countless hours of stress and intense physical and mental activity prepare the Rangers for the complex nature of their assignments around the world.

Rangers are trained to work in harsh conditions, even while hungry, thirsty, and sleep-deprived.

FORCE FACTS "Caliber" has two meanings in weaponry. In small arms such as pistols and rifles, it refers to the diameter of the barrel in inches. But in artillery, it refers to the length of the gun barrel in multiples of its diameter. For example, a 5-inch/50-caliber gun's length is 5 x 50 inches, which is 250 inches (635 cm), or just under 21 feet (6.4 m).

FIREPOWER AND FILMS

Like other Special Forces, the Rangers are overseen by the U.S. Special Operations Command (SOCOM). Operating from three bases in the U.S., the Rangers have the capability to be anywhere in the world within 18 hours. First Battalion is headquartered at Hunter Army Airfield in Savannah, Georgia. Second Battalion is located at Joint Base Lewis–McChord near Tacoma, Washington. Third Battalion and the Regimental Special Troops Battalion (RSTB) are based at Fort Benning, about 250 miles (400 km) west of Hunter Army Airfield.

To carry out their missions, Rangers rely on a wide range of proven firepower. The standard Ranger weapon is the SOPMOD Block II M4A1 carbine. This is a shorter version of the M16 rifle, which entered service more than 50 years ago. Soldiers often add a *foregrip* to it. Side rails allow the weapon to be fitted with a variety of features, including scopes with different magnification levels. Many Rangers attach an M203 grenade launcher under the barrel. It has an effective range of a quarter mile (0.4 km) and often serves as a Ranger's primary weapon at the beginning of an assault. A Ranger may fire several grenades in quick succession before firing his M4.

The standard Ranger sidearm is the Beretta M9 pistol, which is the army's version of the popular civilian model. Some soldiers prefer the .40 S&W Glock. A few Rangers honor their colonial ancestors by carrying a tomahawk on missions, too.

The 1962 film Merrill's Marauders was based on a book by Charlton Ogburn Jr., who served in the unit.

FORCE FACTS After the Civil War, some of Mosby's Rangers formed the Texas Rangers. They became known for fighting American Indians for a number of years before shifting their focus to law enforcement.

Shotguns are used when the soldiers must force their way into locked rooms or buildings. A Ranger places the muzzle of the weapon at an angle just above the door's locking mechanism and pulls the trigger. Another blast or two into the doorjamb, and the way is open. Explosive charges, battering rams, heavy pry bars, or other similar devices may be used to break down heavier doors or external gates. Once the entryway is open, other Rangers pour in with their M4s at the ready.

When more firepower is required, Rangers employ a variety of weapons. One is the M249 light machine gun. Mounted on a *bipod*, it can shoot more than 100 rounds per minute at distances exceeding 2 miles (3.2 km). The MK 47 Striker grenade launcher boasts an advanced fire control system. It can fire a continuous stream of "smart" grenades—which are preprogrammed to burst at a set distance—in addition to conventional projectiles. The M3 "Carl Gustaf" *recoilless* rifle, otherwise known as the Ranger Anti-tank Weapons System (RAWS), can shred "technicals"—as the pickup trucks favored by terrorist groups are known—at distances of more than a mile and a half (2.5 km). The M224 60 mm lightweight mortar can fire up to 30 rounds a minute in short bursts at distances ranging from less than 100 yards (90 m) to more than 2 miles (3.2 km).

Rangers use specific vehicles to get them where they need to go. One is the Ground Mobility Vehicle (GMV), with open sides and rear and no doors. It's a stripped-down, unarmored version of the Humvee light truck. In many cases, Rangers mount a .50 caliber M2 heavy machine gun and/or a 7.62 mm M240 medium machine gun in the back of the GMV for additional protection. In the rugged terrain of countries such as Afghanistan, the GMV often proves too unwieldy for effective use. As a result, Rangers may drive the same Toyota pickup trucks favored by their enemies. This tactic has another advantage. Enemy forces seeing

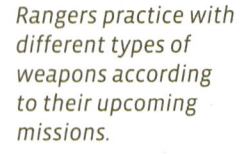

Rangers practice with different types of weapons according to their upcoming missions.

> **FORCE FACTS** Warner Brothers originally wanted Charlton Heston to play Colonel William Darby. But Heston insisted on receiving 5 percent of the film's profits. The studio refused and replaced him with James Garner.

them approach frequently assume that they are friendly. By the time they realize their error, it is too late.

The eight-wheeled M1126 Stryker Infantry Carrier Vehicle is useful in urban environments. Half an inch (1.3 cm) of armor helps protect the vehicle against snipers. The large vehicle is nearly silent, allowing as many as nine Rangers to get as close to their enemies as possible without raising an alarm. This quality makes the vehicle especially useful in nighttime operations. The Stryker can also provide massive firepower to support the Rangers, with weapons such as an M68A2 105 mm cannon, MK19 grenade launcher, and M2 or M240 machine guns. It can even lay down heavy smoke.

Tales of military heroism are always popular among moviegoers, and the elite Army Rangers are no exception. The first Ranger-related film was *Northwest Passage* (1940), starring Spencer Tracy as Major Robert Rogers. The film was a hit both at the box office and among critics. Variety called it "one of the finest epic adventure dramas ever screened." The film has been criticized in recent years as racist in its depiction of American Indians. However, many people maintain that it accurately reflects attitudes at the time of the action.

Warner Brothers released *Darby's Rangers* in 1958. James Garner played Colonel William Darby. While critics praised some of the battle scenes, most felt there was too much emphasis on the characters' love lives. "You might gather from *Darby's Rangers* that the major interest and pursuit of the special combat force of American soldiers that bore that tag in World War II was chasing after women," said one critic. "It's not enough to say that romance is a recurring dis-

Set during the French and Indian War, Northwest Passage *features Spencer Tracy as Robert Rogers.*

The wheeled Stryker Infantry Carrier Vehicle can carry two crew members and nine passengers.

traction in this film," said another. "It would be more accurate to describe the battle scenes as interrupting all the mush."

Four years later, Jeff Chandler starred as General Frank Merrill in *Merrill's Marauders*. The film follows the unit's primary mission through the Burmese jungle to capture the strategic town of Myitkyina. Director Sam Fuller, a World War II combat veteran, wanted to make sure the film was gritty and realistic. In general, critics felt that Fuller had been successful. "It is most effective as a movie which brings to life what it was like as a soldier in the jungles rather than the detailing of the actual mission," said one reviewer. "By its final scene, the film has decisively undermined every illusion of military glory," said another. "It's about a trial of endurance with no reward, no carrot."

Black Hawk Down (2001), a film based on the book by Mark Bowden, dramatized the ill-fated Mogadishu raid. It won Academy Awards for film editing and sound. Director Ridley Scott received an Oscar nomination. Many critics liked the movie. Roger Ebert of the Chicago Sun-Times said, "Films like this are more useful than gung-ho capers like *Behind Enemy Lines*. They help audiences understand and sympathize with the actual experiences of combat troops, instead of trivializing them into entertainments." *Rolling Stone*'s Peter Travers called it "a personal best for producer Jerry Bruckheimer, a triumph for Scott, and a war film of prodigious [remarkably great] power. You will be shaken."

Released in 2005, *The Great Raid* depicts the World War II rescue of prisoners at Cabanatuan in the Philippines. The majority of critics gave the film low marks, mainly for what they felt was its excessive length and subplots. However, Ebert said, "Here is a war movie that understands how wars are actually fought ... it is good to have a film that is not about entertainment for action fans, but about how wars are won with great difficulty, risk, and cost."

Re-enacting the failed 1993 raid of Mogadishu, Black Hawk Down received two Academy Awards.

FORCE FACTS During a break in the filming of Merrill's Marauders, 42-year-old star Jeff Chandler injured his back playing baseball. He underwent surgery but died as a result of complications.

Rogers' Rangers carried tomahawks as well as muskets as they trekked through woods on snowshoes.

RANGERS IN ACTION

PROBABLY THE MOST FAMOUS EXPLOIT OF ROGERS' RANGERS during the French and Indian War was the attack on the village of St. Francis, located near the St. Lawrence River in present-day Canada. The Abenaki Indians living there were accused of torturing two captured British officers. In addition, many people thought that the village served as the home base for raids as far away as Massachusetts. Rogers led more than 200 men on a mission of revenge in mid-September 1759. "Remember the barbarities that have been committed by the enemy's Indian scoundrels on every occasion," said the British general who ordered the raid.

Rogers' Rangers covered about 160 miles (258 km) through dense forests and deep swamps to arrive at their destination. They killed most of the inhabitants, including women and children. Then they burned down the village. One of the Rangers said, "This was I believe the bloodiest scene in all America, our revenge being complete." Now it was the other side's turn to seek revenge. French troops and their Indian allies relentlessly pursued and attacked Rogers' men. The Rangers split into smaller groups to make travel easier and reduce the difficulties of finding food. The half-starved Rangers ate bark, roots, and the flesh off beaver skins. There were even rumors of cannibalism. A number of the men were captured. Some were executed. Rogers himself managed to make it back to safety. Despite those hardships, the exploits served to increase Rogers's fame. The Rangers showed that

Robert Rogers's Rules of Ranging *included tactics that seemed unusual at the time.*

> **FORCE FACTS** The Pointe du Hoc attack forms the basis for parts of several video games, such as *Call of Duty 2*, *G.I. Combat*, and *Company of Heroes*.

they could travel vast distances and launch attacks.

One of the best-known Ranger operations during World War II was the January 1945 rescue of more than 500 Allied prisoners of war (POWs) at the Cabanatuan prison camp in the Philippines. Most of the POWs had been captives for nearly four years. Mistreated and malnourished, many suffered from a variety of diseases. Even worse, the Japanese planned on murdering them rather than allowing them to be freed by the advancing American forces. More than 130 POWs had been massacred at another prison camp the previous month.

To save the men at Cabanatuan, 120 Rangers and about 300 Filipino *guerrillas* set off on a daring mission. As many as 10,000 Japanese soldiers were either at the camp or within a few miles. There was no time to rehearse the mission. Traveling cautiously to avoid roving patrols, the Rangers arrived at the site and hid in nearby underbrush. They would have to crawl over several hundred feet of cleared terrain to reach the camp. Even under cover of darkness, that would be dangerous.

To distract the guards, a plane repeatedly flew low over the camp. The pilot made it sound as if the plane was damaged. The trick worked. The Rangers achieved complete surprise. They killed all the guards and only lost two men. The weakened state of the prisoners meant progress toward safety was slow. Some had to be carried. Nonetheless, aided by American aircraft that protected the route, everyone reached American lines. The prisoners' stories of the brutality they had suffered swept across the U.S. and increased Americans' resolve to defeat the Japanese. General Douglas MacArthur, commander of U.S. forces in the Philippines, said, "No incident of the campaign in the Pacific has given me such satisfaction as the release of the POWs at Cabanatuan. The mission was brilliantly successful."

In recent years, most Ranger missions have been in the Middle East, especially Afghanistan and Iraq. Perhaps the highest-

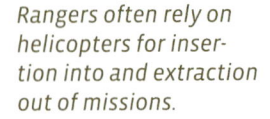

Rangers often rely on helicopters for insertion into and extraction out of missions.

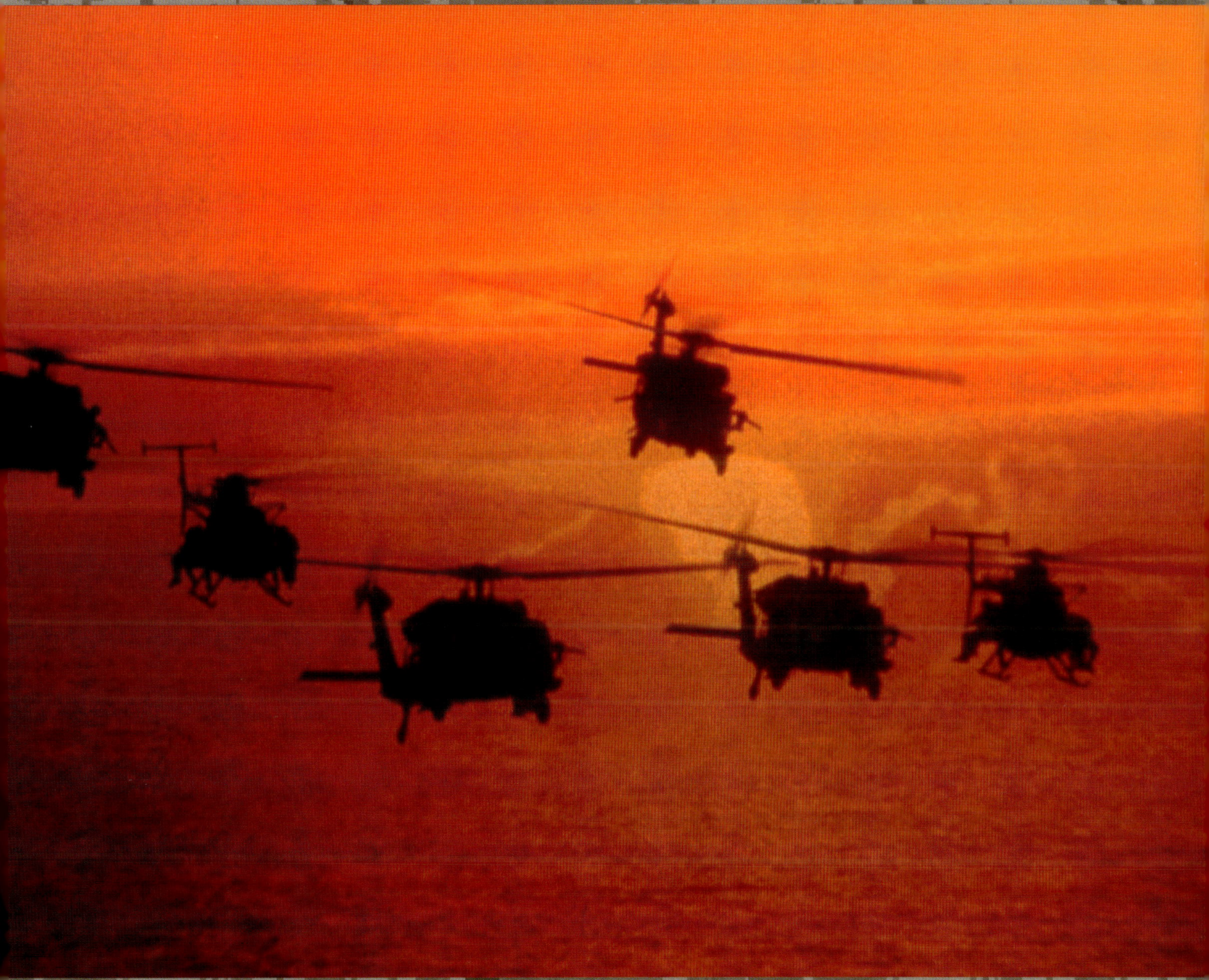

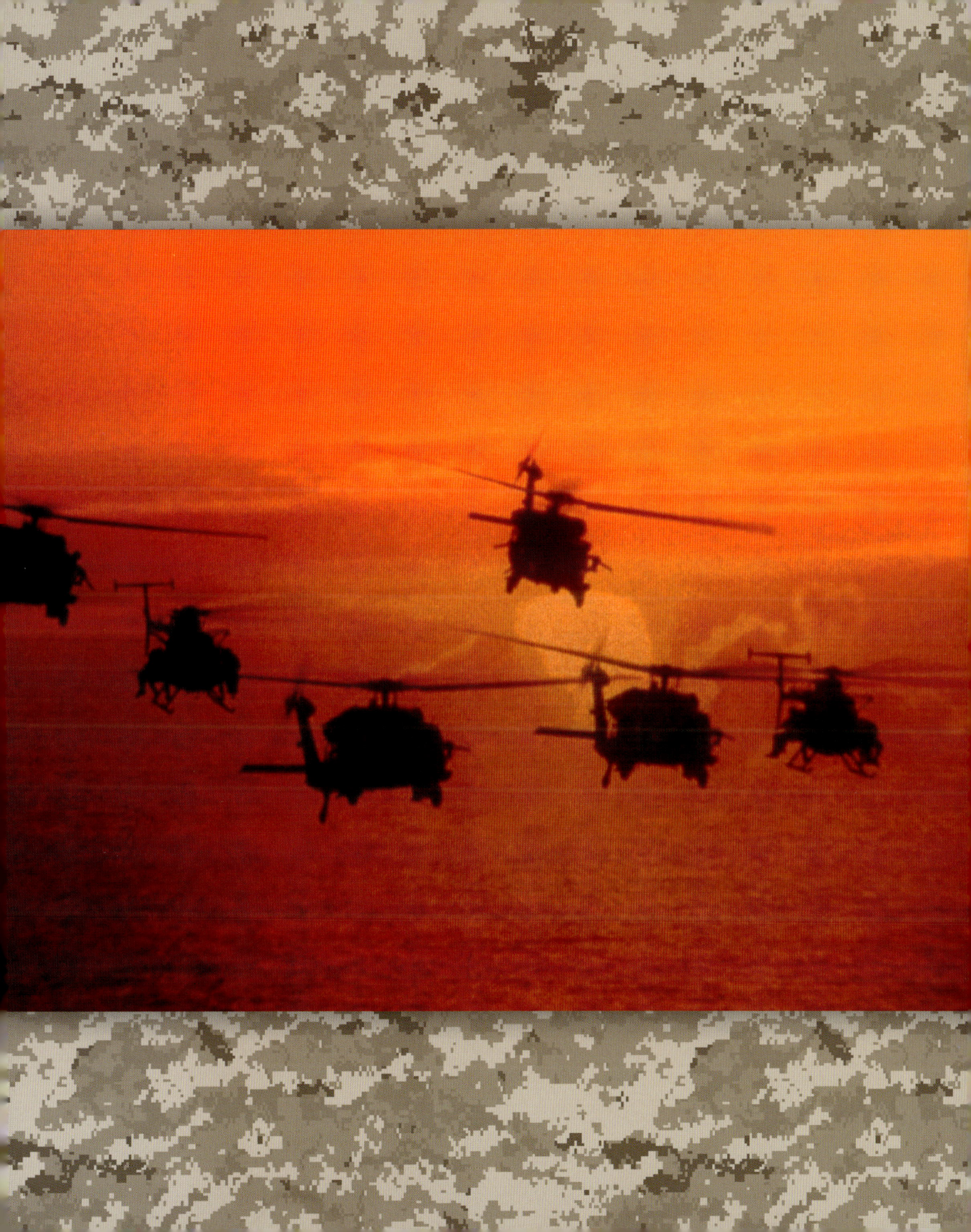

FORCE FACTS According to the Pat Tillman Foundation's website, it "invests in military veterans and their spouses through academic scholarships—building a diverse community of leaders committed to service to others."

profile action involved Pat Tillman. After an outstanding football career at Arizona State University, Tillman starred for the Arizona Cardinals of the National Football League. But in the aftermath of 9/11, he turned down a lucrative multi-year contract and enlisted in the army. He soon joined the Rangers. On April 22, 2004, he was part of a mission to retrieve a Humvee that had broken down in the mountains of southeastern Afghanistan. His platoon was divided into two groups. They had difficulties communicating with each other in the rugged terrain. One group was ambushed in the gathering dusk. The other, which included Tillman, tried to assist them. In the confusion, Tillman was killed. Within a few days, he became a national hero. He was awarded the Silver Star, the nation's third-highest decoration for military heroism. His memorial service was nationally televised. It generated hundreds of news stories throughout the country and led to heightened interest in the army.

Tillman *had* died heroically. He had protected a fellow soldier from withering gunfire. But the shooters hadn't been members of the Taliban. It quickly became apparent to the army that he had been killed by "friendly fire." Fellow Rangers had mistaken him for the enemy in the near-darkness. But high-ranking officers tried to suppress the truth. They ordered men in Tillman's platoon to lie and destroyed vital evidence. Worse yet, they lied to Tillman's family. The truth began emerging shortly after his burial and memorial service. Tillman's brother Kevin said, "The deception surrounding this [Tillman] case was an insult to the family." The incident had one positive outcome, the establishment of the Pat Tillman Foundation. The organization has sponsored hundreds of scholarships since 2004.

Pat Tillman enlisted in the military with his brother Kevin; both completed Ranger training.

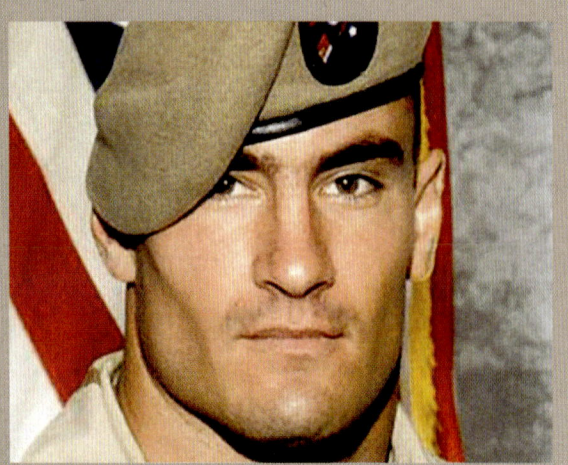

One of the Rangers' most vital missions in the current era is the elimination of high-value targets (HVTs). These are individuals who play key roles in planning and carrying out terror attacks on civilians and U.S. military personnel. On June 24, 2008, Rangers were assigned to take out their highest-ever HVT: Abu Khalaf. He was second-in-command of **al Qaeda in Iraq**. He had been pursued for several years before finally being located in Mosul in a house surrounded by high walls with a heavy steel gate. It was imperative to attack before Khalaf slipped away again.

Leaving their Strykers more than a mile (2 km) away—and thereby losing the support of the vehicles' heavy machine guns if they were discovered—a platoon of Rangers hurried through the shadows toward their objective. They paused while the sniper team crossed rooftops using a 30-foot (9 m) scaling ladder laid flat to get into position to provide covering fire. That took nine minutes. "It feels like it's an eternity when you're sitting on a fairly well-lit street corner at 11:00 P.M. in one of the most hos-

Rangers are called on to complete missions all around the world—many operations take place in the Middle East.

tile cities in Iraq," said one of the Rangers.

When the snipers were in place, Rangers detonated breaching charges that blew down the heavy front gate. The Rangers burst into the house. They immediately encountered a man and woman. They ordered the couple to surrender. The man started to reach inside his vest. He was shot. The woman flung herself onto his body and grabbed at the vest. She, too, was shot. Both had tried to set off an explosive device inside the vest. It would have killed or wounded several Rangers. Moments later, Abu Khalaf came out of hiding. He dashed up a flight of stairs and emerged onto the roof. A sniper shot him. The whole operation took 30 seconds. No Rangers were harmed.

In August 2010, intelligence located another HVT. He was hiding in the Pech Valley of Afghanistan's Kunar Province. An expert in *improvised* explosive devices (IEDs) and small-unit tactics, he was responsible for a number of American *casualties*: A platoon of Rangers helicoptered into the area to take him out. Unfortunately, heavy cloud cover deprived the unit of its *ISR platform*. As a result, the HVT and his men were able to remain undetected as the helicopter landed. When the Rangers emerged onto a rooftop, they came under fire. Although two Rangers were wounded, the rest of the platoon quickly fought back and killed their attackers. Sadly, one Ranger, Chris Wright, died of his wounds.

"It still amazes me to this day how mentally tough Chris Wright was, and the only thing I can think of is that he spent his last moments in strength, confidence, and honor because he was with his brothers to the end, and he knew he had never let anyone down," said fellow Ranger Grant McGarry. Both men embody the Ranger creed. No matter where they are deployed or what types of missions they perform, the motto coined on the bloody beaches of Normandy will always apply: Rangers lead the way!

★ ★ ★

Extensive training enables Rangers to think and act with quickness and confidence in combat situations.

FORCE FACTS Rangers in Afghanistan usually slice up their empty plastic water bottles so that their enemies can't reuse them.

FORCE FACTS The Regimental Reconnaissance Company (RRC) may access its objectives via high-altitude parachute jumps, submarines, small boats, and scuba operations.

Rangers are always prepared to embark on a mission at a moment's notice.

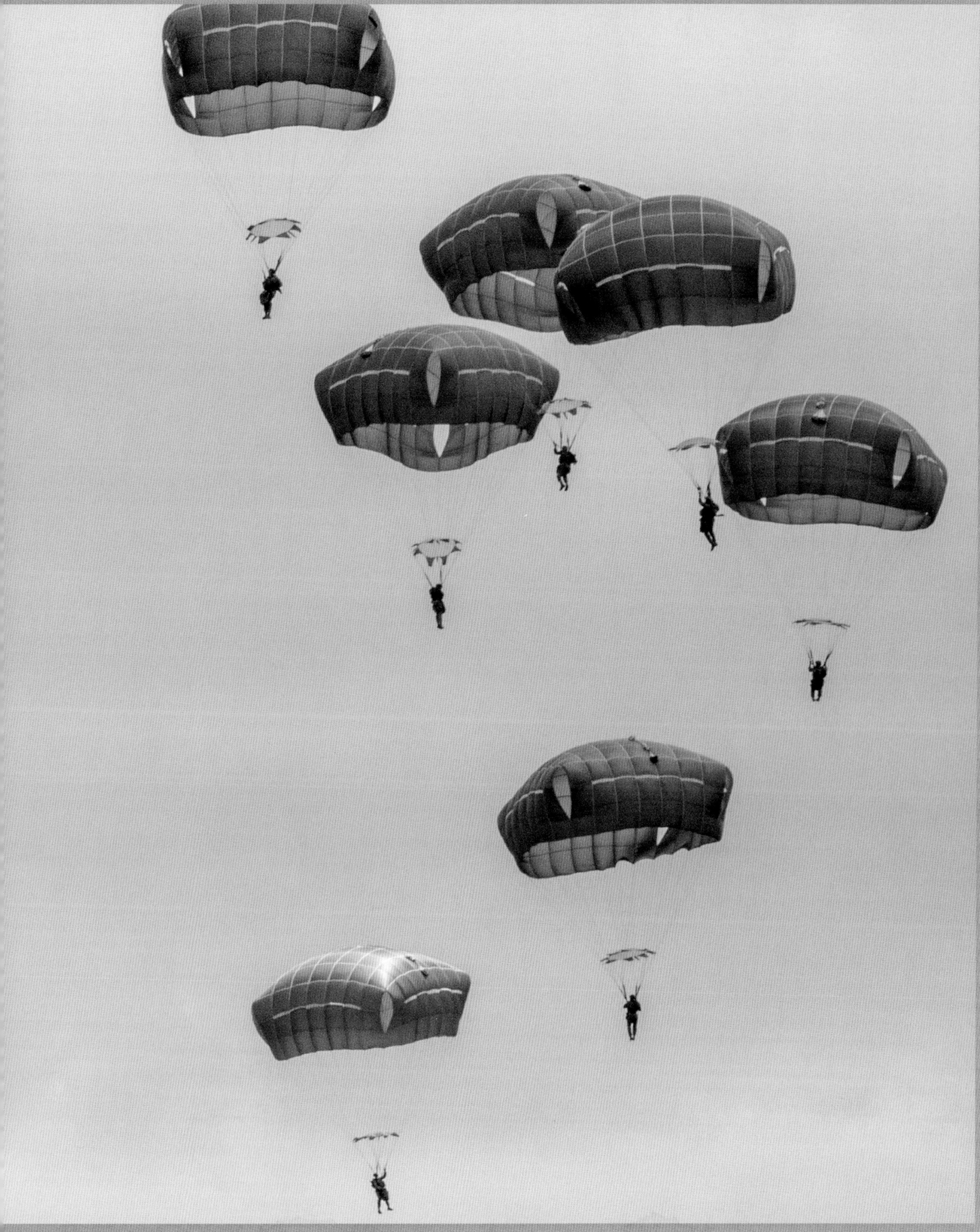

GLOSSARY

al Qaeda in Iraq – a terrorist organization formed in 2004 to fight the U.S. occupation of Iraq

bipod – a support apparatus with two legs

breach – to break through a barrier

casualties – people injured or killed in an accident or a battle

Delta Force – a highly secretive U.S. Army special mission unit

deploy – move personnel into position for military action

foregrip – the handle of a weapon mounted under the front part of the barrel

French and Indian War – a conflict (1754–63) between France and Great Britain for control of North America

grappling hooks – devices with metal hooks attached to a rope and designed to dig into a surface

guerrillas – fighters who aren't part of conventional armed forces

improvised – produced without advance preparation from available materials

intelligence – information about movements and strength of forces of an enemy

ISR platform – an intelligence, surveillance, and reconnaissance system that may include satellites, manned aircraft, and unmanned aircraft such as drones

recoilless – describing lightweight artillery that counteracts and reduces the normal recoil movement of rifles

reconnaissance – a search to gain information, usually conducted in secret

rucksack – a type of backpack made of strong, waterproof material and often associated with the military

small-unit tactics – techniques of dealing with combat situations for platoons and smaller units

Taliban – a fundamentalist Islamic political movement and militia in Afghanistan; noted especially for terror tactics and a repressive attitude toward women

FORCE FACTS When he was 23, Abraham Lincoln joined the Illinois Frontier Guard, one of the mounted ranger units organized during American westward expansion. He never experienced combat.

SELECTED BIBLIOGRAPHY

Bahmanyar, Mir. *Shadow Warriors: A History of the US Army Rangers.* Oxford: Osprey, 2005.

Bohrer, David. *America's Special Forces.* St. Paul, Minn.: MBI, 2002.

Bryant, Russ. *To Be a U.S. Army Ranger.* St. Paul, Minn.: MBI, 2003.

Caraccilo, Dominic J. *Forging a Special Operations Force: The U.S. Army Rangers.* Solihull, UK: Helion, 2015.

Clancy, Tom, with Gen. Carl Stiner and Tony Koltz. *Shadow Warriors: Inside the Special Forces.* New York: Putnam's, 2002.

Frederick, Jim. *Special Ops: The Hidden World of America's Toughest Warriors.* New York: Time Books, 2011.

Neville, Leigh. *US Army Rangers 1989–2015: Panama to Afghanistan.* Oxford: Osprey, 2016.

Tucker, David, and Christopher J. Lamb. *United States Special Operations Forces.* New York: Columbia University Press, 2007.

WEBSITES

Becoming an Army Ranger for the Day
http://wish.org/wishes/wish-stories/i-wish-to-be/eleck-army-ranger#sm.0000ce8vds2v1fccs7g1xgv679bh0

An eight-year-old boy with cystic fibrosis trains with the Rangers for a day through the Make-A-Wish Foundation.

Pointe-Du-Hoc
http://www.worldwar2history.info/D-Day/Pointe-Du-Hoc.html

Read historian Stephen F. Ambrose's detailed account of the Rangers' assault on the German stronghold during the Normandy invasion.

READ MORE

Brush, Jim. *Special Forces.* Mankato, Minn.: Sea-to-Sea, 2012.

Cooper, Jason. *U.S. Special Operations.* Vero Beach, Fla.: Rourke, 2004.

Note: Every effort has been made to ensure that the websites listed above have educational value and that they contain no inappropriate material. However, because of the nature of the Internet, it is impossible to guarantee that these sites will remain active indefinitely or that their contents will not be altered.

INDEX

Abrams, Creighton 15
Afghanistan 16, 28, 36, 38, 40
al Qaeda 39
Burma (Myanmar) 14, 32
Camp Frank D. Merrill 24
Camp James E. Rudder 24
Church, Benjamin 11
Civil War 12, 28
Cold War 14
cooperation with other special forces 14, 16
Cota, Norman 12
Darby, William O. 12, 30
deployment 19, 22, 40
early Rangers history 11, 12, 28, 35, 36
    Mosby's Rangers 12, 28
    Plymouth Colony Militia 11
    Rogers' Rangers 11, 12, 35, 36
Fort Benning, Georgia 20, 23, 27
French and Indian War 11, 35
Grenada 15
headquarters 15, 22, 27
Iraq 16, 36, 39, 40
Jamestown, Virginia 11
Khalaf, Abu 39, 40
Korean War 14
MacArthur, Douglas 36
Marion, Francis 12
McGarry, Grant 40

media portrayals 30, 32, 33, 36
    books 32
    films 30, 32, 33
    video games 36
Merrill, Frank 14, 32
missions 9, 12, 14, 15, 19, 22, 23, 27, 32, 35, 36, 38, 39–40
    direct action 19, 36, 39–40
    and Merrill's Marauders 14, 32
Mosby, John S. 12
motto 12, 40
Noriega, Manuel 16
organizational structure 14, 15, 16, 17, 19, 20, 22, 27, 32, 38, 39, 40, 43
    battalions 14, 15, 16, 17, 19, 20, 22, 27, 43
    companies 14, 16, 19, 22
    personnel 16, 19, 22
    platoons 19, 22, 38, 39, 40
    units 14, 19, 32
Philippines 14, 32, 36
Reagan, Ronald 15
responsibilities 11, 12, 14, 15, 16, 19, 32, 36
    Long Range Reconnaissance Patrol (LRRP) 11, 12, 14
    rescues 14, 15, 19, 32, 36
    scouting 12, 14
Revolutionary War 12
Rogers, Robert 11, 12, 30, 35
Smith, John 11
Somalia 16, 32

Taliban 16, 38
terrorism 16, 28, 39
Tillman, Pat 38
training 7, 14, 16, 19, 20, 22, 23, 24
    Airborne School 19, 20
    Basic Combat Training 20
    and berets 20
    Ranger Assessment and Selection Program 20, 22
    Ranger School 14, 22, 23, 24
    specialized training programs 24
U.S. military operations 15, 16, 32
    Battle of Mogadishu 16, 32
    Eagle Claw 15
    Just Cause 16
    Urgent Fury 15, 16
U.S. Special Operations Command (SOCOM) 27
vehicles 20, 28, 30, 39
Vietnam War 14, 15, 16
War on Terror 16
Washington, George 12
weaponry 8, 20, 22, 23, 24, 27, 28, 30
World War II 8, 9, 12, 14, 32, 36
    Battle of Cisterna 14
    Battle of Myitkyina 14
    Cabanatuam 32, 36
    D-Day 9
    Pointe du Hoc 8, 9, 36
Wright, Chris 40